Chemical Engineering and the Forum (Including Sponsored Technology Workshops)

Held at the 2024 AIChE Annual Meeting

San Diego, California, USA
27-31 October 2024

ISBN: 979-8-3313-1680-8

Printed from e-media with permission by:

Curran Associates, Inc.
57 Morehouse Lane
Red Hook, NY 12571

AIChE
120 Wall Street, FL 23
New York, NY 10005-4020

Phone: (800) 242-4363
Fax: (203) 775-5177

www.aiche.org

Additional copies of this publication are available from:

Curran Associates, Inc.
57 Morehouse Lane
Red Hook, NY 12571 USA
Phone: 845-758-0400
Fax: 845-758-2633
Email: curran@proceedings.com
Web: www.proceedings.com

TABLE OF CONTENTS

Things You Should know to be Effective as an Expert Witness on an Intellectual Property Matter in the Chemical Processing Industry

David A. Rockstraw, Ph. D., P. E. (CA, NM, TX), Inc.

david@davidarockstrawphdpe.com

An ***expert witness*** is a person permitted to testify at trial because of special knowledge or proficiency in a field that is relevant to the case. At all stages in the legal proceedings, on the basis of the evidence as he understands it, an expert witness provides independent assistance to the court and the parties by way of objective unbiased opinion in relation to matters within his expertise and should never assume the role of an advocate.

The expert evidence presented to the court should be, and be seen to be, the independent product of the expert uninfluenced as to form or content by the exigencies of the litigation. An expert witness should state the facts or assumptions upon which his opinion is based. He should not omit to consider material facts which could detract from his concluded opinion. An expert witness should make it clear when a particular question or issue falls outside his expertise. Where an expert is of the opinion that his conclusions are based on inadequate factual information he should say so explicitly.

An expert must remain ready to reconsider his opinion, and if appropriate, to change his mind when he has received new information or has considered the opinion of the other expert, and should do so at the earliest opportunity.

A ***fact witness*** has personal knowledge of events pertaining to the case and can testify as to things they have personally observed or witnessed. Fact witnesses may not offer opinions.

Chemical Engineers are often called upon to offer opinions in areas that include (not exhaustive):

- Chemical/Materials Process Design;
- Explosions/Fires Cause & Origin/Insurance;
- Contracts;
- Chemical/Materials Characterization;
- Regulatory, Safety, & Hazards;
- Intellectual Property/Trade Secrets;
- Chemical/Materials Process Design, Simulation, Analysis;
- Explosions/Fires Cause & Origin;
- Materials/Chemical; and
- Safety, Hazards, Compliance.

Intellectual property (IP) litigations can include both patent infringement and trade secret litigations. The former may take place in venues that include: U.S. District Courts; U.S. International Trade Commission (ITC); The United States Patent & Trademark Office (IPRs, CBMs); The European Patent Office (EPO); and The World Intellectual Property Organization (WIPO); whereas Trade Secret Litigation is heard by U.S. District Courts and State Courts.

Intellectual Property & Patent Law is defined in Article I, Section 8, the U.S. Constitution:

> Congress shall have power . . . To promote the progress of science and useful arts, by securing for limited times to authors and inventors the exclusive right to their respective writings and discoveries.

This law is promulgated through the Patent Act (35 U.S. Code § 102 of Federal Regulations), which contains five parts:

- Part I – United States Patent & Trademark Office
- Part II - Patentability of Inventions and Grants of Patents
- Part III – Patents and Protection of Patent Rights
- Part IV – Patent Cooperation Treaty
- Part V – Hague Agreement Concerning International Registration of Industrial Designs

Infringement of a patent refers to the unauthorized manufacture, sale, or use of a literal copy of a patented invention or an invention that performs substantially the same function in substantially the same manner as a patented invention. (***doctrine of equivalents***). For the element to be "equivalent" it must exhibit only insubstantial changes from that which is claimed. A test to determine equivalence is whether the element in the accused product performs substantially the same function in substantially the same way to achieve substantially the same results as the claimed limitation.

Assessing infringement of a U.S. Patent requires 2 steps: (1) construe the ***claim terms*** of the patent claims to understand what meaning they would have to a person of ordinary skill in the art (***POSITA***); and (2) performing an infringement analysis requires a determination of whether the accused product, or its use, contains elements that correspond to each and every limitation present in an asserted claim.

A ***claim*** is the critical defining element of a patent, and the primary subject of examination by the US Patent Office. It is the claims of a patent that are compared to the ***prior art*** references to determine novelty and non-obviousness. Claims appear as a series of numbered sentences at the end of a patent, defining the extent of legal protection conferred. Each claim consists of an introductory segment (a "preamble"), followed by one or more phrases or clauses ("claim elements") separated by commas or semicolons.

Each claim element (or portion thereof) may be referred to as a claim ***limitation*** since it adds another necessary element to the scope of the invention, thereby limiting the class of infringing devices, methods, etc., to those having that limitation.

In order to exclude someone from using a patented invention, the patent owner must demonstrate that what the potential infringer is using falls within the scope of at least one claim of a patent (i.e., that the infringing device, method, etc., has all the limitations of at least one claim of the patent).

Loosely, ***prior art*** refers to all information that is publicly available before someone claims to invent something. Patents are only allowed for things that are novel and nonobvious. If the prior art contained a description of the supposed invention, it usually cannot be novel. If the prior art contained enough information that the invention would have been obvious to someone of ordinary skill in the relevant field, a patent also cannot issue.

A *POSITA* (person of Ordinary Skill in the Art) is a hypothetical person considered to have the normal skills and knowledge in a particular technical field (an "art"), without being a genius. Factors that may be considered in determining level of ordinary skill in the art:

- the education and level of skill of the inventors;
- the types of problems encountered in the prior art;
- the prior art solutions to those problems;
- the rapidity with which innovations are made;
- the sophistication of the technology; and
- the education level and experience of active workers in the field.

In a given case, every factor may not be present, and one or more factors may predominate. To assess the level of ordinary skill in the art, in addition to reading the patent, one can consider:

- the types of problems encountered in the field of the invention,
- the prior solutions to those problems found in prior art references,
- the sophistication of the technology, and
- the level of education of active workers in the field.

A example of a POSITA Definition may be as follows:

> "A POSITA at the time of invention of the Asserted Patents would have had a Bachelor's degree in Chemical Engineering or Chemistry with at least 2 to 3 years of experience in the chemical engineering industry, including experience with operating and/or implementing chemical production facilities that involve the phase equilibrium of liquids and vapors and partially miscible liquids. Additional education might substitute for some of the experience, and substantial experience might substitute for some of the educational background. "

Discovery is a phase of pretrial procedure in a lawsuit in which each party, through the law of civil procedure, obtains evidence from other parties by means of methods of discovery that include interrogatories and requests for production of documents.

Discovery can be obtained from nonparties using subpoenas. When a discovery request is objected to, the requesting party may seek the assistance of the court by filing a motion to compel discovery. Conversely, a party or nonparty resisting discovery can seek the assistance of the court by filing a motion for a protective order

Claim Construction is another phase of pretrial procedure during which claim terms are interpreted according to their plain & ordinary meaning to a POSITA. Where the parties have offered different constructions claim terms, the Court releases its constructions for these terms.

The PTO determines the scope of claims in patent applications not solely on the basis of the claim language, but upon giving claims their broadest reasonable construction "in light of the specification as it would be interpreted by one of ordinary skill in the art." (Fed. Cir. 2004)

The transitional phrases ***comprising***, ***consisting essentially of*** and ***consisting of*** define the scope of a claim with respect to what unrecited additional components or steps, if any, are excluded from the scope of the claim.

- ***comprising*** is synonymous with including, containing, or characterized by, is inclusive or open-ended and does not exclude additional, unrecited elements or method steps.
- ***consisting of*** excludes any element, step, or ingredient not specified in the claim.
- ***consisting essentially of*** limits the scope of a claim to the specified materials or steps and those that do not materially affect the basic and novel characteristic(s) of the claimed invention.

The determination of what is or is not excluded by a transitional phrase must be made on a case-by-case basis considering the facts of each case.

Assessing infringement of a U.S. Patent requires 2 steps: (1) construe the claim terms of the patent claims to understand what meaning they would have to a person of ordinary skill in the art; and (2) performing an infringement analysis requires a determination of whether the accused product, or its use, contains elements that correspond to each and every limitation present in an asserted claim.

A patent issued by the United States Patent and Trademark Office ("USPTO") is presumed to be valid. Due to this presumption, the USPTO is presumed to have established the validity of the patented claims over the prior art. The burden to prove invalidity rests with the Defendants. Invalidity is shown if the Defendants demonstrate invalidity (patent was ***obvious*** §103 or ***anticipated*** §102 by prior art) by clear and convincing evidence, which is higher than the preponderance of evidence standard used for findings of infringement.

The Supreme Court recognizes several rationales for combining references or modifying a reference to show obviousness of claimed subject matter. Some of these include:

- combining prior art elements according to known methods to yield predictable results;
- simple substitution of one known element for another to obtain predictable results;
- use of a known technique to improve a similar device (method, product) in the same way;
- applying a known technique to a known device (method, or product) ready for improvement to yield predictable results;
- choosing from a finite number of identified, predictable solutions, with a reasonable expectation of success; and
- some teaching, suggestion, or motivation in the prior art that would have led a PHOSITA to modify the prior art reference or to combine prior art reference teachings to arrive at the claimed invention.

Secondary considerations for obviousness include whether:

- the invention was commercially successful as a result of the merits of the claimed invention (rather than a result of design needs or market-pressure, advertising, or similar);
- the invention satisfied a long-felt need;
- others had tried and failed to make the invention;
- others invented the invention at roughly the same time;
- others copied the invention;
- changes or related technologies or market needs contemporaneous with the invention;
- the invention achieved unexpected results;
- others in the field praised the invention;
- persons having ordinary skill in the art of the invention expressed surprise or disbelief regarding the invention;
- others sought or obtained rights to the patent from the patent holder; and
- the inventor proceeded contrary to accepted wisdom in the field.

When considering issues of obviousness, it is necessary to determine the scope and content of the prior art; ascertain the differences between the prior art and the claims at issue; resolve the level of ordinary skill in the pertinent art; and consider any evidence of secondary considerations.

A patent may be ***unenforceable*** due to issues of ***inequitable conduct*** (when a patent applicant makes affirmative misrepresentation of material fact, failed to disclose material information, or submitted false material information intended to deceive the PTO) or ***patent misuse*** (a patent owner may exploit a patent but may not use it to acquire a monopoly not embraced in the patent).

Patent misuse occurs when the patent owner has imposed conditions that derive their force from the patent, which impermissibly broadens the physical or temporal scope of the patent grant with anticompetitive effect (e.g., refusing to license a patent to someone in particular; licensing a patent under the condition that the buyer also buy another product or patent license; or using one's own invention in a way that would make other people infringe upon a patent).

The Uniform Trade Secrets Act (USTA) defines a **trade secret** as information (including a formula, pattern, compilation, program, device, method, technique, or process) that:

- derives independent economic value, actual or potential, from not being generally known to, and not being readily ascertainable by proper means by, other persons who can obtain economic value from its disclosure or use; and
- is the subject of efforts that are reasonable under the circumstances to maintain its secrecy.

Prior to the UTSA, Restatement of Torts (1939) defined 6 factors generally considered in determining whether information constitutes a trade secret:

1. extent to which the information is known outside the claimant's business;
2. extent to which it is known by employees and others in the business;
3. extent of measures taken to guard secrecy of the information;
4. value of the information to the business and its competitors;
5. amount of effort or money expended in developing the information; and
6. ease or difficulty information could be properly acquired or duplicated by others.

The IP litigation process takes on average ~ 2½ years, passing through four distinct phases:

- Phase I - filing of the complaint to the case management conference (4-8 months)
    - Initial case analysis, responding to complaint, initial discovery, initial witness interviews, early settlement and licensing discussions, motion to transfer (or stay) case and case management conference ("CMC").
    - 50% of Cases in Texas or Delaware Settle in First 8-11 Months of Litigation
    - 5-10% of fees, <5% of costs
- Phase II - scheduling conference to claim construction hearing (8-10 months)
    - From CMC through end of Markman hearings, including ongoing discovery and analysis, submission of Markman briefs, technical tutorial and Markman before the judge
    - Markman Hearing - a pretrial hearing in a U.S. District Court during which a judge examines evidence from all parties on the appropriate meanings of relevant key words used in a patent claim, when patent infringement is alleged by a plaintiff. It is also known as a "Claim Construction Hearing"
    - Roughly 90% of Cases Settle Before Claim Construction
    - 15-20% of fees, 5-15% of costs
- Phase III - Claim construction hearing to end of expert discovery (6-8 Months)
    - From Markman hearing to completion of fact and expert discovery, summary judgment briefing and hearing,
    - 95-97% of Cases Settle Before Trial
    - 30-40% of fees, 30-35% of costs
- Phase IV - Pre-trial and Trial (3-5 Months)

- judge may establish some basic rules for the case proceedings;
  - judge will typically set a schedule for the trial, and for other pretrial matters;
  - parties may argue other what evidence should and should not be included at trial;
  - parties may argue over whether certain witnesses should be used at trial;
  - parties may request a change of venue (location of court where trial is held);
  - either party's attorney may make motions for dismissal or summary judgment, asking judge to dismiss the matter entirely or rule on points of law favorable to their client;
  - the issues to be decided at trial may be identified and narrow at pretrial hearing; and
  - parties may attempt to settle the matter without going to trial (civil case) or work out a plea bargain (criminal case).
- Trial
- 30-40% of fees, 45-50% of costs

To work in the legal sector as an expert witness, a chemical engineer needs proper academic credentials, appropriate experience in the area relevant to the litigation, a solid reputation of ethical behavior, strong communication skills, and a degree of salesmanship.

Author Biography

Dr. David A. Rockstraw, P. E. is an experienced expert witness in the field of chemical engineering, chemical process analysis/forensics, intellectual property & trade secrets, and chemical safety/hygiene matters, among other areas involving chemical and materials phenomena. He has testified in trials, arbitrations, and depositions dozens of times in six different countries, and has the experience of over 70 past or current expert engagements, as well as over 30 technical consultant engagements. Dr. Rockstraw is the former Robert Davis Distinguished Professor and Academic Department Head of Chemical Engineering, Interim Academic Department Head of Mechanical & Aerospace Engineering and Director of Brewery Engineering at New Mexico State University, with over a decade of experience in the chemical process industry (DuPont, Conoco, Ethyl, Kraft) and national laboratories (Los Alamos, Sandia). Rockstraw holds a Professional Engineering license in TX, NM, and CA.

192b An Introduction to Intellectual Property for Established and Inspiring Start-Ups and Entrepreneurs

Michael Stephenson

This presentation will provide aspiring and established start-ups and entrepreneurs with basic--but crucial---knowledge regarding intellectual property, particularly patents and trade secrets. Discussion topics will also include best practices and tips for start-ups and entrepreneurs.

386a Democratizing Modeling, Simulations, and Computational Data Sciences to Accelerate Process Development and Manufacturing in Pharma and Chemical Industry

Neda Nazemifard

Workshop Format:
Duration: 2 hours
Structure:
Short Presentation (15 minutes):
The workshop will kick off with a brief presentation highlighting the current challenges facing pharma industry and the landscape of modeling and simulations and their potential benefits in process development and manufacturing.

Panel Discussion (90 minutes):
Following the presentation, a panel discussion will feature experts from various sectors:
Pharmaceutical Industry: Leaders who have integrated modeling and simulations into their development pipelines.
Academia: Researchers contributing to advancements in computational sciences and data-driven methodologies.
Modeling Software Companies: Innovators providing cutting-edge simulation tools that power digital transformation.
The discussion will focus on key topics such as:
How can we democratize access to advanced modeling and simulation tools for non-modelers in our organizations?
Current obstacles in implementing these technologies in process development, including the skills gap, high costs, and resistance to change.
Collaboration models between industry, academia, and software companies to accelerate innovation and address key challenges.
Future trends and opportunities in leveraging data science, AI, and digital twins to further enhance process development and manufacturing.

739a Assessing Adsorbent Materials for Applications in Direct Air Capture (DAC) Utilizing Breakthrough and Static Adsorption Measurements

Julian Hungerford

Breakthrough and static adsorption measurements are important for the development materials used in direct air capture (DAC) of carbon dioxide. These techniques are particularly valuable for evaluating material performance in low CO2 concentrations and in the presence of water. By utilizing these analytical methods, researchers can reduce the development time of new materials and processes for DAC, by measuring surface texture, adsorption capacity, and selectivity.

740a Digital Threads for Data and Process Traceability and Reproducibility in Drug Discovery

Tara Madhyastha

This presentation demonstrates the impact of workflow automation in chemical engineering, highlighting how life sciences companies automate molecular and quantum dynamics analysis in pharmaceutical development. Using Rescale and Nextflow, a Fortune 500 life sciences customer saved time by automating manual data management, reduced duplication and re-work, and ultimately sped up the drug development timeline. By making complex workflows easier and more efficient, the customer reduces R&D costs and delivers new medicines to market faster, with greater confidence in their efficacy for patients.

Rescale works with leading life sciences companies to build digital threads that transform pharmaceutical development by connecting teams and automating complex scientific workflows. Digital threads play a central role in life sciences R&D, linking each step of the drug discovery process, from modeling simulation to data analysis, making it faster and easier for global teams from different disciplines to collaborate seamlessly from anywhere. Digital threads also provide a trace of the specific steps involved in producing consistent results, with modularity for teams to use their best-fit software and hardware technologies.

Using Rescale and Nextflow to build computational workflows for multi-step analyses, the example life sciences company provides their scientists and researchers with a digital thread of processes, data, and tools so they focus on answering key research questions. This approach improves accessibility and reproducibility of complex workflows critical for faster turnarounds in collaborative research, while enabling them to swap in any preferred commercial, open-source, and custom technologies. With Rescale and Nextflow, any collaborator can quickly automate chains of analysis from a variety of applications and teams involved have visibility of their process and data for greater speed and confidence in decision-making.

741a Purdue Center for Operations and Optimization in Process Systems

David E. Bernal Neira

The Purdue Center for Operations and Optimization in Process Systems (COOPS) is at the forefront of transforming the field of process systems engineering. By leveraging cutting-edge research and fostering innovation, COOPS aims to enhance the efficiency, sustainability, and safety of various industries.
Our center focuses on:
- conducting state-of-the-art research combining modeling and theoretical studies to solve complex systems engineering problems
- engaging students at all levels (undergraduate, MS, and PhD) in hands-on research and specialized courses to prepare the next generation of leaders in PSE
- providing expert consultation and innovative solutions for industrial challenges, ensuring our research to have practical impact
- collaborate with other faculty and departments in Purdue University to promote interdisciplinary research and education

Our mission is to drive innovation and excellence in various critical areas, including process operations and control, scheduling, smart manufacturing, crystallization, predictive maintenance, energy systems optimization, life cycle analysis, process design, supply chain optimization, artificial intelligence, and quantum computing.

AUTHOR INDEX